AF687868

Molecular Storms and Self-Assembly

by

Ayele Teklemariam

RoseDog Books
PITTSBURGH, PENNSYLVANIA 15238

The contents of this work, including, but not limited to, the accuracy of events, people, and places depicted; opinions expressed; permission to use previously published materials included; and any advice given or actions advocated are solely the responsibility of the author, who assumes all liability for said work and indemnifies the publisher against any claims stemming from publication of the work.

All Rights Reserved
Copyright © 2025 by Ayele Teklemariam

No part of this book may be reproduced or transmitted, downloaded, distributed, reverse engineered, or stored in or introduced into any information storage and retrieval system, in any form or by any means, including photocopying and recording, whether electronic or mechanical, now known or hereinafter invented without permission in writing from the publisher.

Rosedog Books
585 Alpha Drive
Pittsburgh, PA 15238
Visit our website at *www.rosedogbooks.com*

ISBN: 979-8-89341-039-6
eISBN: 979-8-89341-537-7

I owe a debt of gratitude and deep appreciation to my beloved parents, who had a high enough belief in and regard for education to have sent us all, my siblings and me, to school. My siblings, particularly my older siblings, made it possible for me to get an early-enough acquaintance with reading and writing by making their textbooks, reading books, and notes available when I needed them, and their stimulating dialogue and conversations to date on every occasion the family gets a chance to convene. This is the foundation on which I anchor the pole that supports the house I call "me." I say thank you, and I am grateful.

Ayele Teklemariam
Also @
https://soundcloud.com/newsayele-teklemariam
http://cybrntn1.wordpress.com
http://cybernatian.net
http://ezinearticles.com/?expert=Ayele_Teklemariam

Molecular Storms and Self-Assembly

A molecular storm is a term used to describe the constant and random motion of molecules and atoms at the nanoscale. This motion is caused by the thermal energy of the system, and it is responsible for many of the fundamental processes of life, such as diffusion, chemical reactions, and protein folding.

Molecular storms are also thought to play a role in the formation of complex structures such as crystals and living cells. For example, when molecules collide with each other in a random fashion, they can sometimes form ordered structures by chance. This process is known as self-assembly, and it is thought to be responsible for the formation of many of the complex structures we see in the natural world.

In addition to their role in the formation of complex structures, molecular storms also play a role in the destruction of existing structures. For example, molecular storms can cause proteins to unfold and DNA to break apart. This can lead to cell death and other problems.

Overall, molecular storms are a fundamental part of the nanoscale world. They are responsible for many of the processes that are essential for life, but they can also be destructive.

Here are some examples of molecular storms in the real world:

The diffusion of oxygen and carbon dioxide in the lungs

The chemical reactions that take place in cells

The folding of proteins into their three-dimensional shapes

The self-assembly of viruses and other biological structures

The destruction of cells by cancer and other diseases

Molecular storms are also being studied by scientists and engineers to develop new technologies such as nanobots and self-assembling materials. The concept of using nano-robots to trigger self-assembly in molecular storms is an intriguing one with potential applications in various fields. Here's a hypothetical scenario of how this might work:

Imagine a swarm of minuscule nano-robots, each equipped with specialized sensors and actuators, dispersed within a carefully controlled molecular storm environment. These nano-robots would be programmed to monitor the chaotic dynamics of the storm, identifying and manipulating specific molecules to initiate and guide self-assembly processes.

By strategically introducing specific molecules or manipulating the local environment around them, the nano-robots could influence the interactions between molecules, coaxing them into forming desired structures. For instance, they could introduce seed molecules that act as nucleation sites for crystal growth, or they could manipulate the electrostatic forces between molecules to promote specific orientations.

The nano-robots could also act as dynamic scaffolds, providing temporary support or guidance to developing structures until they mature enough to become self-supporting. They could even communicate with each other, coordinating their actions to achieve larger, more complex structures.

This hypothetical scenario highlights the potential of combining nano-robots with the inherent chaos of molecular storms to achieve controlled self-assembly of sophisticated structures. Such technology could have far-reaching implications in fields such as materials science, nanotechnology, and medicine.

Some potential applications of this technology:

Materials science: self-assembled structures with tailored properties could be created for various applications, such as lightweight and strong materials for construction, biocompatible materials for implants, or highly efficient catalysts for energy production.

Nanotechnology: self-assembled nanostructures could be used to create electronic devices with enhanced performance, develop novel sensors with unprecedented sensitivity, or design nanomachines with advanced functionalities.

Medicine: self-assembled structures could be used to deliver drugs or therapies directly to diseased cells, construct tissue scaffolds for regenerative medicine, or engineer nano-carriers for targeted drug delivery.

While the concept is still in the realm of science fiction, the potential benefits of harnessing molecular storms for self-assembly using nano-robots are significant. As nanotechnology and our understanding of molecular dynamics continue to advance, this hypothetical scenario may one day become a reality.

Here are some examples of molecular storms in the real world:

- The diffusion of oxygen and carbon dioxide in the lungs
- The chemical reactions that take place in cells
- The folding of proteins into their three-dimensional shapes
- The self-assembly of viruses and other biological structures
- The destruction of cells by cancer and other diseases

Molecular storms are also being studied by scientists and engineers to develop new technologies, such as nanobots and self-assembling materials.

Although there are some who claim that the necessity of an initial prime mover or creator is preempted by the process of molecular storm-caused self-assembly, the distinct differences between biological self-generation and regeneration distinguishes itself from any other forms of self-assembly and invalidate their arguments.

In the context of origin-of-life research, the terms "prime mover" and "self-assembly" are often used to describe two fundamental aspects of the transition from non-living matter to living systems.

Prime mover refers to the initial source of energy or information that kickstarted the process of self-organization and complexity leading to the first lifeforms. The exact nature of the prime mover remains a mystery, but various hypotheses have been proposed, including

Energy sources: lightning strikes, meteor impacts, or hydrothermal vents could have provided the energy necessary for the synthesis of organic molecules from simpler inorganic compounds.

External forces: the shockwaves from a nearby supernova or the gravitational pull of a passing comet could have triggered chemical reactions leading to the formation of life's building blocks.

Chemical gradients: the interplay between different chemical environments, such as the interface between water and land, could have facilitated the formation of complex molecules.

Self-assembly refers to the ability of molecules to spontaneously organize themselves into ordered structures without external guidance or control. This process is essential for the formation of biological structures such as cell membranes, proteins, and nucleic acids.

The interplay between the prime mover and self-assembly is crucial for understanding the origin of life. The prime mover provided the initial energy or information that set the stage for self-assembly, while self-assembly allowed for the formation of increasingly complex structures that eventually gave rise to life.

Researchers continue to investigate the potential mechanisms by which these two processes could have interacted to bring about the emergence of life. Understanding these mechanisms is a critical step in unraveling one of the most profound mysteries of science of the origin of life on Earth. Finding the beginning might take more than getting to the beginning, but getting to place time and condition past the beginning is not only challenging, rather impossible. Perhaps what laid beyond the beginning is a black hole, permanently locked or perhaps destroyed by the beginning itself, beyond recovery that all efforts could possibly be in vain.

Looking from our vantage point out to the beginning, we hit an impenetrable wall of the beginning, beyond where we can neither observe nor perceive as we hit an information blackout.

We can, of course, hypothesize and theorize any and all, but there is no vantage point for us to stand past the beginning to measure, observe, and prove any of the theories or hypotheses about the beginning.

It feels as though the creation or synthesis of synthetic biology has recreated the initial conditions for life's emergence, but the fact remains that life emerged from infinite conditions, infinite possibilities and impossibilities, that one possibility had nothing to do with the specific initial state of the life we know; besides, synthetic biology could possibly be sued for copyright violation by nature for being a parrot-ish mimicry of natural biology. It is after all a chemical synthesis based on knowledge gained by closely and intensely studying nature and its ways. Synthetic biology to me seems a simulation of natural biology devoid of any original authentic creative act.

Molecular storm-driven self-assembly and biological self-generation and regeneration are two distinct processes that share some similarities but also have fundamental differences.

Molecular storm-driven self-assembly refers to the spontaneous organization of molecules into complex and ordered structures under the chaotic conditions of a molecular storm. These storms are characterized by rapid and irreversible chemical transformations, which provide the energy and driving force for self-assembly. The resulting structures can be of various sizes and complexities, ranging from simple nanoscale structures to intricate networks and architectures.

Biological self-generation and regeneration are processes that occur in living organisms and are guided by genetic information. Self-generation refers to the ability of an organism to produce new individuals of its own kind, while regeneration refers to the ability of an organism to repair or replace damaged or lost tissues. Both processes rely on the precise orchestration of biochemical pathways and cellular processes.

Here's a table summarizing the key differences between the two processes:

Feature	Molecular storm-driven self assembly	Biological self-generation and regeneration
Driving force	Chaos and random collisions	Genetic information and biochemical pathways
Level of organization	Molecular	Cellular and organism
Control	Spontaneous	Guided
Products	Complex and ordered structures	New individuals or repaired tissues
Examples	Formation of snowflakes, crystals, embryo development, wound healing, and membranes	Tissue regeneration

Despite their differences, both molecular storm-driven self-assembly and biological self-generation and regeneration are fascinating examples of the ability of matter to organize itself into complex and functional structures. These processes are essential for the existence of life and continue to inspire scientists and engineers in their quest to create new materials and technologies.

Synthetic life and biological life are two distinct categories of life forms, with different origins, characteristics, and potential implications.

Biological life is the form of life that has evolved naturally on Earth over billions of years. It is characterized by its ability to self-replicate, grow, adapt, and evolve. Biological organisms are composed of cells, which are the basic units of life. Cells contain DNA, the genetic material that carries the instructions for building and maintaining an organism.

Synthetic life, on the other hand, is created in a laboratory by scientists. It is not produced through natural evolutionary processes but rather through the manipulation of non-living matter to create life-like structures or systems. Synthetic life forms may exhibit some of the same characteristics as biological life, such as self-replication or metabolism, but they lack the complexity and adaptability of natural organisms.

Here is a table summarizing the key differences between synthetic life and biological life:

Feature	Biological life	Synthetic life
Origin	Natural evolution	Laboratory synthesis
Composition	Cells containing DNA	Non-living matter
Complexity	Highly complex and diverse	Varies from simple to complex
Adaptability	Can evolve and adapt to new environments	Limited adaptability
Self-replication	Ability to replicate itself	May or may not be able to self-replicate

Metabolism—ability to carry out metabolic processes; may or may not have a metabolism

The creation of synthetic life is a relatively new field of research, and scientists are still exploring the potential of this technology. However, there are some potential implications of synthetic life that are worth considering:

Medical applications: Synthetic life could be used to develop new drugs, therapies, or diagnostic tools. For example, synthetic cells could be used to deliver drugs directly to diseased cells, or they could be used to create models of human diseases for research purposes.

Environmental applications: synthetic life could be used to clean up pollution, produce renewable energy, or develop new materials with sustainable properties.

Industrial applications: synthetic life could be used to produce new materials, improve manufacturing processes, or create new forms of biotechnology.

The development of synthetic life raises a number of ethical and societal questions that need to be carefully considered. For example, who should own and control synthetic life forms? What are the potential risks of creating life that is not bound by natural evolutionary processes? What are the implications for our understanding of what it means to be alive?

As the field of synthetic biology continues to advance, it is important to have open and informed discussions about the potential benefits and risks of this technology. By carefully considering the ethical and societal implications of synthetic life, we can ensure that this technology is used responsibly and for the benefit of humanity. The life of every biological life form and species is a novel written by the script of DNA synthetic life; in comparison is a rudimentary children's bedside story that has yet to traverse long ways to be in par with biological life's exotic and complex novels.

Molecular storms are also being studied by scientists and engineers to develop new technologies, such as nanobots and self-assembling materials.

Thermodynamic Equilibrium

Thermodynamic equilibrium is a state of a system in which all macroscopic properties are unchanging in time. This means that there are no net macroscopic flows of matter or energy within the system or between the system and its surroundings, or the system is closed.

Thermodynamic equilibrium is a fundamental concept in thermodynamics, and it is used to describe a wide range of systems, from simple gases to complex living organisms.

One of the key features of thermodynamic equilibrium is that it is a state of maximum entropy. Entropy is a measure of disorder, and it always increases over time in isolated systems. This means that a system in thermodynamic equilibrium is as disordered as it can be, given its constraints.

Another key feature of thermodynamic equilibrium is that it is in a state of minimum free energy. Free energy is a measure of the amount of energy that is available to do work. A system in thermodynamic equilibrium has the minimum amount of free energy that it can have, given its constraints.

Thermodynamic equilibrium is an important concept in many different areas of science and engineering. For example, it is used to design heat engines, refrigerators, and other thermodynamic systems. It is also used to study the behavior of living organisms, as they are constantly trying to maintain a state of thermodynamic equilibrium.

Some examples of thermodynamic equilibrium:

A gas in a sealed container at a constant temperature and pressure is in thermodynamic equilibrium.

A liquid in a beaker at a constant temperature is in thermodynamic equilibrium. A living cell is in thermodynamic equilibrium, although it is constantly exchanging matter and energy with its surroundings.

Thermodynamic equilibrium is a complex topic, but it is an important one to understand for anyone who wants to study thermodynamics or engineering.

Free energy is a measure of the amount of energy that is available to do work. It is calculated by subtracting the enthalpy (heat content) of a system from its Gibbs energy (total energy).

Free energy can be positive, negative, or zero. A positive free energy means that the system has more energy than it needs to do work. A negative free energy means that the system has less energy than it needs to do work, and it must absorb energy from its surroundings in order to do work. A zero free energy means that the system is at equilibrium and cannot do any work.

Things move because they have free energy. Free energy is used to drive processes such as diffusion, chemical reactions, and muscle contraction. For example, when a gas diffuses from a high-concentration region to a low-concentration region, it is doing work because it is moving against the concentration gradient. This work is done by the free energy of the gas.

Chemical reactions also require free energy to occur. For example, when glucose is converted to carbon dioxide and water in the presence of oxygen, the reaction releases free energy. This free energy is used to do work, such as powering muscle contraction or generating ATP.

Muscle contraction also requires free energy. When a muscle contracts, the actin and myosin filaments slide past each other. This requires the muscle to use free energy to break and reform bonds between the actin and myosin filaments. Overall, free energy is essential for things to move. It is the driving force behind many of the processes that occur in the natural world.

Here are some specific examples of how free energy is used to drive things to move:

Diffusion: Gases and liquids diffuse from high-concentration regions to low-concentration regions. This is driven by the free energy of the system.

Chemical reactions: chemical reactions that release free energy, such as combustion, can be used to power machines and other devices.

Muscle contraction: muscles contract by using free energy to break and reform bonds between the actin and myosin filaments.

Plant growth: plants use free energy from sunlight to drive photosynthesis, which produces the carbohydrates that they need to grow.

Animal locomotion: animals use free energy from food to power their muscles and move around.

Free energy is a fundamental concept in physics and chemistry, and it is essential for understanding how things move in the natural world, particularly in the nano- and microworlds. Overall, thermodynamics provides a framework for understanding the relationship between energy, entropy, and the emergence of chaos or order in complex systems. It highlights the delicate balance between energy flow, entropy production, and self-organization that underlies the behavior and dynamics of these systems.

Peter M. Hoffmann, in his book *Life's Ratchet*, delves into the mystery of life from the perspective of nanotechnology and molecular machines. He explores how seemingly purposeful motions within cells and organisms actually arise from microscopic chaos, termed the "molecular storm."

Hoffmann argues that life's secret lies in the nanoscale world, where specialized molecules act as miniature ratchets. These "molecular machines" exploit the random energy of water molecules to create order and purposeful activity within cells, nano and microscales, where the rules of quantum mechanics are in observance.

Erwin Schrödinger's Famous Thought Experiment

Schrödinger's cat is a thought experiment devised by Austrian physicist Erwin Schrödinger in 1935. The experiment is designed to illustrate the problems with the Copenhagen interpretation of quantum mechanics.

The debate over Schrödinger's cat and the interpretation of quantum mechanics is ongoing. It is a complex and challenging topic, but it is also one that is essential for understanding the nature of reality.

Erwin Schrödinger's famous thought experiment, known as Schrödinger's cat, is a way of illustrating the strange and paradoxical nature of quantum mechanics. In the experiment, a cat is placed in a box with a radioactive atom. If the atom decays, it will trigger a mechanism that releases a poison gas, killing the cat. However, until the box is opened, and the atom is observed, it is in a superposition of both decayed and non-decayed states. This means that the cat is also in a superposition of both alive and dead states.

Schrödinger's cat experiment highlights the fact that classical physics, which is based on the idea of objective reality, is not adequate to describe the behavior of matter at the quantum level. In quantum mechanics, particles can exist in multiple states at the same time, and the act of observation collapses the superposition and forces the particle to choose a definite state.

This has profound implications for our understanding of the nature of reality. If reality is super-positional, then it means that there is no such thing as an objective reality. Instead, reality is constantly in flux, and it is only when we observe it that it collapses into a definite state. In other words, reality is realized only by an observer, or unobserved reality is no more.

Understanding Schrödinger's dilemma is essential for understanding the nature of reality because it forces us to confront the limitations of our classical understanding of the world. It also opens up new possibilities for our understanding of consciousness and the role of the observer in shaping reality.

Here are some specific ways in which understanding Schrödinger's dilemma can help us to understand the nature of reality:

It challenges our classical understanding of reality as objective and deterministic. In classical physics, particles have definite positions and momenta, and their behavior is governed by deterministic laws. In quantum mechanics, particles can exist in multiple states at the same time, and their behavior is probabilistic and a result of stochastic mathematics. This suggests that reality may not be as objective or deterministic as we thought.

It highlights the importance of the observer. In classical physics, the observer is separate from the observed system. In quantum mechanics, the observer plays an active role in shaping the outcome of an experiment. This suggests that consciousness may play a more fundamental role in reality than we thought.

It opens up new possibilities for our understanding of space and time. In classical physics, space and time are absolute entities. In quantum mechanics, space and time may be more fluid and relational. This suggests that our understanding of space and time may need to be revised.

Overall, understanding Schrödinger's dilemma is essential for understanding the nature of reality because it forces us to confront the limitations of our classical understanding of the world and opens up new possibilities for our understanding of consciousness, space, and time.

Molecular storms and stochastic mathematics are two seemingly unrelated fields that actually have a lot in common. Stochastic mathematics is the study of random or probabilistic processes, while molecular storms are sudden, chaotic bursts of activity within cells.

Both stochastic mathematics and molecular storms can be modeled using mathematical equations. For example, stochastic differential equations can be used to model the behavior of molecular storm particles, while Markov chains can be used to model the transitions between different states of a molecular storm.

A Markov chain is a mathematical model used to describe a sequence of events where the probability of each event depends only on the state of the previous event. It's like a simple prediction machine, using the present to estimate the future.

Imagine a weather model that only considers yesterday's weather to predict today's. That's a Markov chain!

Here are the key aspects of a Markov chain:

States: These are the possible conditions or situations the system can be in. In the weather example, states could be sunny, rainy, or cloudy.

Transitions: These are the probabilities of moving from one state to another. For instance, the probability of rain after a sunny day might be lower than after a cloudy day.

Memoryless property: This is the crucial part. The future only depends on the present state, not the entire history that led to it.

Markov chains are used in various fields, including:

Finance: predicting stock market movements.

Natural language processing: generating text, like the words you're reading now!

Recommendation systems: suggesting products or movies you might like based on your past choices.

Games: creating AI opponents that can adapt to your playing style.

Here's an image to help visualize a Markov chain:

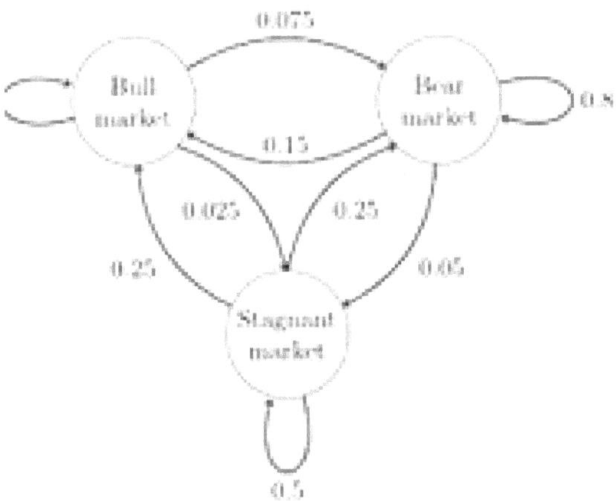

The dots represent states, and the arrows show the transition probabilities between them. For instance, the arrow from "sunny" to "rainy" might be shorter than the arrow from "rainy" to "sunny," reflecting the higher chance of rain after a rainy day.

Stochastic mathematics can also be used to study the effects of molecular storms on cells. For example, researchers can use stochastic models to predict how molecular storms affect gene expression and protein synthesis.

Overall, stochastic mathematics is a powerful tool that can be used to study and understand molecular storms. By using stochastic models, researchers can gain insights into the behavior of these complex systems and develop new strategies for treating diseases that are caused by molecular storms.

Stochastic modeling can be used to study the emergence of order from molecular storms or chaos. This is because stochastic models can capture the inherent randomness of these systems, as well as the interactions between individual molecules.

One example of how stochastic modeling has been used to study order formation is in the context of protein folding. Protein folding is a complex process in which a long chain of amino acids folds into a three-dimensional structure. This structure is essential for the protein to function properly.

Stochastic models have been used to show that protein folding can be driven by the random collisions of individual amino acids. These collisions can cause the amino acids to stick together, forming the initial structure of the protein. Over time, this structure can then refine itself until the protein reaches its final, folded state.

Stochastic modeling has also been used to study order formation in other systems, such as the formation of patterns in sand dunes. Sand dunes are formed by the wind blowing sand across a surface. The wind causes the sand to move in a random fashion, which can lead to the formation of patterns such as ripples and ridges.

Stochastic models have shown that the formation of these patterns can be explained by the random interactions between individual sand grains. The models have also shown that the patterns are sensitive to the wind speed and direction, as well as the properties of the sand itself.

Stochastic modeling is a powerful tool that can be used to study order formation in complex systems. By capturing the inherent randomness of these systems, as well as the interactions between individual molecules, stochastic models can provide valuable insights into the processes that govern order formation.

If the universe was spread uniformly evenly with whatever it is made of and makes it, it wouldn't be hard to imagine where we all would be, if we would be at all. In other words, everything and anything at all that had what we call a material and objective reality, it owes its existence and continuance to the mere fact of uneven distribution of the stuff of the universe all across the board.

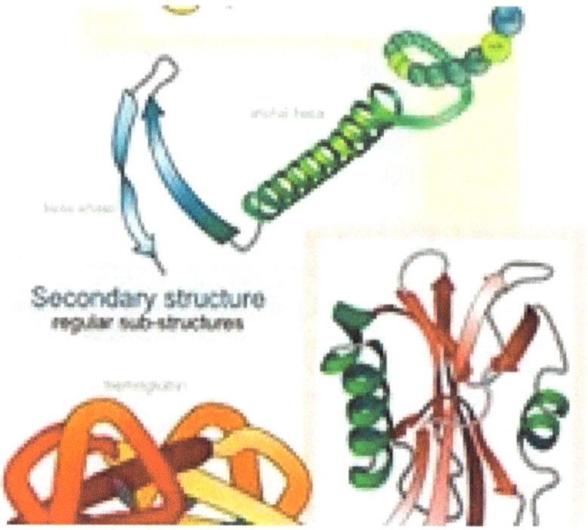

As the unevenness in the universe is characterized by uneven distribution of the stuff of the universe, the universe itself is in turn an entity that may be a product and continuum of an unevenness. The unevenness that gave us the galaxies, planets, stars, and planets is the one that gave us the continents, countries, and regions uneven, unique, and different. Uneven distribution did not only spread the stuff that made anything and everything but made them unique and different; the same uniqueness and difference that transcends all existence of material reality bar none. That is so, as suggested by quantum mechanics, that the existence of two or more similar quantum states is not allowed by any material reality at the same time, though there seems an exception to quantum-entangled entities, which even then are separated by space, which renders the exception as none exception. As that is a subject of more intense study and not the object of this writing, suffice it to say that much.

The subject of uneven distribution and what it means to our daily lives and how it has affected us and how it will continue to shape and mold us to be better and fit for the lives and life forms to follow. We are as all life is, a product sculpted, forged and formed, shaped, reshaped, ground and refined, tuned and returned in response, and by the force of uneven distribution of the stuff of the universe that is instrumental to the reality of diverse environment. In life, more so than any reality, in the universe is this diversity more pronounced, both for a reason and a vital one.

Diverse life, which came about as direct result of the unevenness that defines the known universe more than any of the other attributes, exhibits a unique reverse cycle that seems and perhaps truthfully accentuates and necessitates more of diversity and unevenness. Diversity is the progenitor of specialization in its crude form and more refined and complex forms that necessitates and perpetrates cooperation and interdependence to hold the system standing in place and continue. Pulling out of the specialized member entities is like pulling out one of the pillars of the system or structure that renders the stability of the system suspect. When we get to the race of humanity, nation and tribe, if nothing else, the diversity in specialty gets even more pronounced than anytime and anywhere else.

Perhaps it is profoundly important to note that specialization and cooperation of variances in specialty is foundational and is manifested in our earlier beginnings right from the time pluripotent cells proliferation to specialized cells of the different organs set to function in the unevenness of the environment of the inner bodies of life forms. The emergence of specialized cells from pluripotent cells, a process called differentiation, is like a grand symphony, where a single starting note (the pluripotent cell) unfolds into a complex and beautiful composition of diverse instruments (specialized cells).

Here's how it plays out:

The pluripotent cell, often a ball of unspecialized potential, sits in a niche rich with chemical signals and cues. These signals act like the conductor's baton, instructing the cell on which path to take.

Inside the cell, the genetic code lies dormant, like a vast library of unread books. Each gene is a recipe for a specific protein, and the cell carefully chooses which ones to activate, like a chef picking ingredients.

As the cell receives signals, it begins to express certain genes and silence others. This fine-tuning of gene activity creates a unique genetic melody within the cell, biasing it towards a particular lineage (ectoderm, mesoderm, or endoderm).

Imagine the conductor tapping out a rhythm on the drum, signaling the percussion section to take the lead. This is like the cell activating genes related to a specific germ layer.

Within each germ layer, the melody becomes more complex. The cell delves deeper into specific pathways, expressing genes for different cell types. It's like the percussion section dividing into timpani, cymbals, and drums, each with their own special sound.

Regulatory proteins, like musical virtuosos, modify chromatin (the packaging of DNA) to lock in these gene expression patterns and prevent unwanted melodies from playing.

As the differentiation program progresses, the cell's physical structure transforms. Organelles like muscles or neurons start to assemble, and the once-smooth, spherical shape becomes specialized, like a violin gaining its distinct curves.

The cell loses its pluripotency, becoming committed to its chosen fate. It's like the musician dedicating themself to mastering their instrument, no longer able to play all the others.

From a single pluripotent cell, a multitude of specialized cells emerge, each with its own unique form and function. They work together in harmony, like the orchestra playing its final piece, to build and maintain the complex tissues and organs that make us who we are.

This is just a simplified overview, of course. The differentiation process is far more intricate, with countless players and feedback loops involved. But, hopefully, it gives you a sense of the awe-inspiring choreography that unfolds as pluripotent cells give rise to the beautiful unity in diversity of life.

Scientists are still unraveling the secrets of this symphony, but understanding differentiation holds immense promise for regenerative medicine, disease treatment, and even creating artificial organs. The music of life continues to play, and pluripotent cells remain the key conductors, shaping the melody of our existence.

The same argument holds true of the melodies played by our inner and outer organs variances and unity in diversity whereby each group of specialized cells function in tandem with every other specialized group of cells for the whole wellbeing of the living animal or plant. The melody and the orchestra continue as we step up the ladder and see the interactions between the different species of plants and animals with only a higher degree of complexity and stormy intricacies. As we discover the hidden mysteries of life, what we have come to realize is none other than the organization and specialization of cells into varied specialties to form the whole of our complex system. The depth of complexity and specialization increases and decreases as the complexity of the task of our group of specialized cells, and no wonder brain cells excel all others by all measures reiterated.

Bacteria, especially Cyanobacteria, have played a crucial role in the evolution and success of life on Earth. Cyanobacteria, also known as blue-green algae, were the first organisms to carry out oxygenic photosynthesis, a process that produces oxygen as a byproduct. This was a significant development as it led to the oxygenation of Earth's atmosphere, creating an environment suitable for the emergence and survival of aerobic organisms.

The special ability of Cyanobacteria to produce oxygen through photosynthesis not only transformed the composition of the atmosphere but also provided a new source of energy for other organisms. Oxygen became a vital component for the development of more complex life forms, including plants and animals. The oxygen produced by Cyanobacteria allowed for the evolution of aerobic respiration, a more efficient process for extracting energy from organic compounds.

Furthermore, Cyanobacteria played a crucial role in the formation of the Earth's ozone layer. The oxygen they released into the atmosphere reacted with ultraviolet radiation, forming ozone molecules that created a protective layer in the stratosphere. This ozone layer shielded life on Earth from harmful UV radiation, enabling the colonization of land by organisms.

In addition to their contributions to the atmosphere and the ozone layer, Cyanobacteria also played a role in shaping Earth's ecosystems. They were instrumental in the formation of stromatolites, layered structures composed of sediment and microbial mats. These stromatolites, formed by the activities of Cyanobacteria, were some of the earliest evidence of life on Earth and helped create habitats for other organisms.

Overall, Cyanobacteria's ability to carry out oxygenic photosynthesis, produce oxygen, and shape Earth's ecosystems has had a profound impact on the evolution and success of life on our planet. They laid the foundation for the development of more complex organisms and created an environment conducive to the diversity, specialty organized complexity, and interdependence of life we see today.

Thermodynamics plays a significant role in understanding the concepts of chaos and order in complex systems. The laws of thermodynamics govern the behavior of energy and its transformations within a system, providing insights into the flow of energy and the emergence of order and chaos.

The second law of thermodynamics, specifically the concept of entropy, is particularly relevant to the understanding of chaos and order. Entropy is a measure of the disorder or randomness within a system. According to the second law, the entropy of an isolated system tends to increase over time, leading to a natural tendency towards disorder or chaos.

In complex systems, such as living organisms or ecosystems, the interplay between energy flow and entropy production is crucial. Energy inputs from the environment are utilized by the system to maintain order and complexity, but this process also generates entropy. The system must constantly balance the creation of order with the increase in entropy to maintain its stability and functionality.

The concept of self-organization is another important aspect related to thermodynamics and the emergence of order in complex systems. Self-organization refers to the spontaneous formation of patterns or structures within a system without external control or intervention. This phenomenon arises from the interactions and exchanges of energy and matter within the system, guided by the laws of thermodynamics.

Molecular Storms and Their Intricacy

A phenomenon where numerous molecules interact in a complex and dynamic fashion is referred to as a "molecular storm" in a number of scientific domains. This can happen in a number of situations, such as:

Cellular processes: To perform vital tasks like gene expression, protein synthesis, and cell signaling, molecular storms within cells might entail the interaction of proteins, DNA, RNA, and other molecules. These storms frequently exhibit strong activity and quick changes.

Chemical processes: Numerous molecules interacting with one another and a sequence of intermediate steps are common in complex chemical reactions. This can be compared to a storm of reactants and products whirling around, where the complex interaction of different elements determines the final result.

Stellar environments: Extreme heat and radiation can split apart and rearrange molecules, creating a chaotic "molecular storm" of different chemical species in the harsh environment of stars. This may play a role in the genesis of new elements and the development of stars.

Prebiotic soup: It is thought that a varied mixture of organic molecules interacted through a variety of mechanisms in the "molecular soup" that was the early Earth's atmosphere. Life is believed to have finally emerged as a result of this intricate interaction.

Complicated ideas are intimately related to the idea of molecular storms. Numerous interacting components define complex systems, which also frequently display emergent traits that cannot be predicted from the individual components alone. This is why studying molecular storms is both interesting and difficult.

Molecular storms are specifically related to complexity in the following ways:

Non-linearity: Small changes in one molecule can have significant and unexpected effects on others because of the non-linear interactions between molecules in a storm. Emergent behavior that is challenging to predict or model may result from this.

Self-organization: Individual molecules may spontaneously arrange themselves into more complex shapes or patterns during molecular storms. Biological processes such as the creation of cellular membranes or the assembly of protein complexes are examples of this.

Adaptability: Molecular storms and other complex systems are frequently able to adjust to changes in their surroundings. This enables them to endure and change with the environment.

Emergence: Molecular storms have the ability to produce features that do not exist in individual molecules, as was previously indicated. This is a defining characteristic of complex systems and can result in strange and unusual behaviors.

Numerous scientific fields are actively conducting research on molecular storms. The goal of studying these storms' behavior is to help scientists better understand complex systems and the building blocks of life.

Citations

Kondepudi, D., & Prigogine, I. (2014). *Modern Thermodynamics: From Heat Engines to Dissipative Structures.* John Wiley & Sons.

Schneider, E. D., & Kay, J. J. (1994). *Life as a Manifestation of the Second Law of Thermodynamics. Mathematical and Computer Modelling*, 19 (6–8), 25–48.

Nicolis, G., & Prigogine, I. (1977). *Self-Organization in Non-equilibrium Systems: From Dissipative Structures to Order Through Fluctuations.* John Wiley & Sons.

Book: Eric D. Schneider and David E. Osterloh's "Molecular Storms: The Physics of Stars, Cells, and the Origin of Life"

William Bialek"s article titled ""Sailing the Ocean of Complexity: Lessons from the Physics-Biology Interface."" Santa Fe Institute website.

Schaefer, B. May 26, 2015. *Chemical Industry Use of Natural Products*. Springer.

Schaefer, B. May 26, 2015. *Chemical Industry Use of Natural Products*. Springer. The Thakkar, A. J. June 12, 1973. *Chemistry Using Computers*. Springer.

Hoffmann, P. M. (2012, October 30). *Life's Ratchet*. Hachette UK. http://books.google.ie/books?id=zpRVDgAAQBAJ&printsec=frontcover&dq=Life %27s+ratchet&hl=&cd=1&source=gbs_api

Lane, N. (2010, June 14). *Life Ascending: The Ten Great Inventions of Evolution*. W. W. Norton & Company.

www.ingramcontent.com/pod-product-compliance
Lightning Source LLC
LaVergne TN
LVHW071220220926
848486LV00037B/273
* 9 7 9 8 8 9 3 4 1 0 3 9 6 *